EDMOND HALLEY

GREAT ASTRONOMERS

ROBERT STAWELL BALL

GIFT *Certificate*

TO:

FROM:

DATE:

Would you like to buy a copy of
EDMOND HALLEY?

PLEASE VISIT:
http://www.diamondbooks.ca

EDMOND HALLEY

GREAT ASTRONOMERS

BY

ROBERT STAWELL BALL

D.Sc, LL.D., F.R.S.

(July 1, 1840 – November 25, 1913)

LOWNDEAN PROFESSOR OF ASTRONOMY AND GEOMETRY
IN THE UNIVERSITY OF CAMBRIDGE

DIAMOND BOOKS™

www.diamondbooks.ca

TORONTO, CANADA – 2017

DIAMOND BOOKS - CANADA

DIAMOND BOOKS

Toronto, ON, CANADA
http://www.diamondbooks.ca

BIBLIOGRAPHIC INFORMATION

'EDMOND HALLEY'
Is an extract or a chapter from the book
"GREAT ASTRONOMERS" by Robert Stawell Ball.

FIRST PUBLISHED IN: 1896.

PUBLISHED IN CANADA

Published in Canada by DIAMOND BOOKS - CANADA, an imprint of DIAMOND PUBLISHERS - http://www.diamondpublishers.com

FIRST EDITION: MARCH, 2017.

PAPERBACK EDITION : ISBN: 978-1-988357-48-5

PRINTED IN CANADA

PREFACE
TO THE ORIGINAL EDITION

It has been my object in these pages to present the life of each astronomer in such detail as to enable the reader to realise in some degree the man's character and surroundings; and I have endeavoured to indicate as clearly as circumstances would permit the main features of the discoveries by which he has become known.

There are many types of astronomers—from the stargazer who merely watches the heavens, to the abstract mathematician who merely works at his desk; it has, consequently, been necessary in the case of some lives to adopt a very different treatment from that which seemed suitable for others.

ROBERT STAWELL BALL.

The Observatory, Cambridge.
October, 1895.

NOTE: The full-length of 'Preface' can be found in its original edition of the book "GREAT ASTRONOMERS".

INTRODUCTION

Of all the natural sciences there is not one which offers such sublime objects to the attention of the inquirer as does the science of astronomy. From the earliest ages the study of the stars has exercised the same fascination as it possesses at the present day. Among the most primitive peoples, the movements of the sun, the moon, and the stars commanded attention from their supposed influence on human affairs. The practical utilities of astronomy were also obvious in primeval times. Maxims of extreme antiquity show how the avocations of the husbandman are to be guided by the movements of the heavenly bodies. The positions of the stars indicated the time to plough, and the time to sow. To the mariner who was seeking a way across thc trackless ocean, the heavenly bodies .offered the only reliable marks by which his path could be guided. There was, accordingly, a stimulus both from intellectual curiosity and from practical necessity to follow the movements of the stars. Thus began a search for the causes of the ever-varying phenomena which the heavens display.

- ROBERT STAWELL BALL

Many of the earliest discoveries are indeed prehistoric. The great diurnal movement of the heavens, and the annual revolution of the sun, seem to have been known in times far more ancient than those to which any human monuments can be referred. The acuteness of the early observers enabled them to single out the more important of the wanderers which we now call planets. They saw that the star-like objects, Jupiter, Saturn, and Mars, with the more conspicuous Venus, constituted a class of bodies wholly distinct from the fixed stars among which their movements lay, and to which they bear such a superficial resemblance. But the penetration of the early astronomers went even further, for they recognized that Mercury also belongs to the same group, though this particular object is seen so rarely. It would seem that eclipses and other phenomena were observed at Babylon from a very remote period, while the most ancient records of celestial observations that we possess are to be found in the Chinese annals. The study of astronomy, in the sense in which we understand the word, may be said to have commenced under the reign of the Ptolemies at Alexandria. The most famous name in the science of this period is that of Hipparchus, who lived and worked at Rhodes about the year 160 B.C. It was his splendid investigations that first wrought the observed facts into a coherent branch of knowledge. He recognized the primary obligation which lies on the student of the heavens to compile as complete an inventory as possible of the objects which are there to be found. Hipparchus accordingly commenced by undertaking, on a small scale, a task exactly similar to that on which modern astronomers, with all available appliances of meridian circles, and photographic telescopes, are constantly engaged at the present day. He compiled a catalogue of the principal fixed stars, which is of special value to astronomers, as being the earliest work of its kind which has been handed down. He also studied the movements of the sun and the moon, and framed theories to account for the incessant changes which he saw in progress. He found a much more difficult problem in his attempt to interpret satisfactorily the complicated movements of the planets. With the view of constructing a theory which should give some coherent account of the subject, he made many observations of the places of these wandering stars. How great were the advances which Hipparchus accomplished may be appreciated if we reflect that, as a preliminary task to his more purely astronomical labours, he had to invent that branch of mathematical science by which alone the problems he proposed could be solved. It was for this purpose that he devised the indispensable method of calculation which we now know so well as trigonometry. Without the aid rendered by this beautiful art it would have been impossible for any n ally important advance in astronomical calculation to have been effected.

But the discovery which shows, beyond all others, that Hipparchus possessed one of the master-minds of all time was the detection of that remarkable celestial movement known as the precession of the equinoxes. The inquiry which conducted to this discovery involved a most profound investigation, especially when it is remembered that in the days of Hipparchus the means of observation of the heavenly bodies were only of the rudest description, and the available observations of earlier dates were extremely scanty. We can but look with astonishment on the genius of the man who, in spite of such difficulties, was able to detect such a phenomenon as the precession, and to exhibit its actual magnitude. I shall endeavour to explain the nature of this singular celestial movement, for it may be said to offer the first instance in the history of science in which we find that combination of accurate observation with skilful interpretation, of which, in the subsequent development of astronomy, we have so many splendid examples.

The word equinox implies the condition that the night is equal to the day. To a resident on the equator the night is no doubt equal to the day at all times in the year, but to one who lives on any other part of the earth, in either hemisphere, the night and the day are not generally equal. There is, however, one occasion in spring, and another in autumn, on which the day and the night are each twelve hours at all places on the earth. When the night and day are equal in spring, the point which the sun occupies on. the heavens is termed the vernal equinox. There is similarly another point in which the sun is situated at the time of the autumnal equinox. In any investigation of the celestial movements the positions of these two equinoxes on the heavens are of primary importance, and Hipparchus, with the instinct of genius, perceived their significance, and commenced to study them. It will be understood that we can always define the position of a point on the sky with reference to the surrounding stars. No doubt we do not see the stars near the sun when the sun is shining, but they are there nevertheless. The ingenuity of Hipparchus enabled him to determine the positions of each of the two equinoxes relatively to the stars which lie in its immediate vicinity. After examination of the celestial places of these points at different periods, he was led to the conclusion that each equinox was moving relatively to the stars, though that movement was so slow that twenty five thousand years would necessarily elapse before a complete circuit of the heavens was accomplished. Hipparchus traced out this phenomenon, and established it on an impregnable basis, so that all astronomers have ever since recognised the precession of the equinoxes) as one of the fundamental facts of astronomy. Not until nearly two thousand years after Hipparchus had made this splendid discovery was the explanation of its cause given by Newton.

From the days of Hipparchus down to the present hour the science of astronomy has steadily grown. One great observer after another has appeared from time to time, to reveal some new phenomenon with regard to the celestial bodies or their movements, while from time to time one commanding intellect after another has arisen to explain the true import of the facts of observations. The history of astronomy thus becomes inseparable from the history of the great men to whose labours its development is due. In the ensuing chapters we have endeavoured to sketch the lives and the work of the great philosophers, by whose labours the science of astronomy has been created.

We shall commence with Ptolemy, who, after the foundations of the science had been laid by Hipparchus, gave to astronomy the form in which it was taught throughout the Middle Ages. We shall next see the mighty revolution in our conceptions of the universe which are associated with the name of Copernicus. We then pass to those periods illumined by the genius of Galileo and Newton, and afterwards we shall trace the careers of other more recent discoverers, by whose industry and genius the boundaries of human knowledge have been so greatly extended. Our history will be brought down late enough to include some of the illustrious astronomers who laboured in the generation which has just passed away.

- ROBERT STAWELL BALL

TABLE OF CONTENTS

CHAPTERS

**An original Chapter from the book: "GREAT ASTRONOMERS", by Robert Stawell Ball.*

CHAPTER

8

GREAT ASTRONOMERS

Edmond Halley

(November 8, 1656 – January 25, 1742)

Edmond Halley

(November 8, 1656 – January 25, 1742)

CHAPTER 8

GREAT ASTRONOMERS: EDMOND HALLEY

ISAAC NEWTON was just fourteen years of age when the birth of Edmund Halley, who was destined in after years to become Newton's warmly attached friend, and one of his most illustrious scientific contemporaries, took place. There can be little doubt that the fame as an astronomer which Halley ultimately acquired, great as it certainly was, would have been even greater still had it not been somewhat impaired by the misfortune that he had to shine in the same sky as that which was illumined by the unparalleled genius of Newton.

- ROBERT STAWELL BALL

Edmund Halley was born at Haggerston, in the Parish of St. Leonard's, Shoreditch, on October 29th, 1656. His father, who bore the same name as his famous son, was a soap-boiler in Winchester Street, London, and he had conducted his business with such success that he accumulated an ample fortune. I have been unable to obtain more than a very few particulars with respect to the early life of the future astronomer. It would, however, appear that from boyhood he showed considerable aptitude for the acquisition of various kinds of learning, and he also had some capacity for mechanical invention. Halley seems to have received a sound education at St. Paul's School, then under the care of Dr. Thomas Gale.

Here the young philosopher rapidly distanced his competitors in the various branches of ordinary school instruction. His superiority was, however, most conspicuous in mathematical studies, and, as a natural development of such tastes, we learn that by the time he had left school he had already made good progress in astronomy. At the age of seventeen he was entered as a commoner at Queen's College, Oxford, and the reputation that he brought with him to the University may be inferred from the remark of the writer of "Athenae Oxonienses," that "Halley came to Oxford with skill in Latin, Greek, and Hebrew, and such a knowledge of geometry as to make a complete dial." Though his studies were thus of a somewhat multifarious nature, yet it is plain that from the first his most favourite pursuit was astronomy.

His earliest efforts in practical observation were connected with an eclipse which he observed from his father's house in Winchester Street. It also appears that he had studied theoretical branches of astronomy so far as to be conversant with the application of mathematics to somewhat abstruse problems.

Up to the time of Kepler, philosophers had assumed almost as an axiom that the heavenly bodies must revolve in circles, and that the motion of the planet around the orbit which it described must be uniform. We have already seen how that great philosopher, after very persevering labour, succeeded in prov-ing that the orbits of the planets were not circles, but that they were ellipses of small eccentricity. Kepler was, however, unable to shake himself free from the prevailing notion that the angular motion of the planet ought to be of an uniform character around some point. He had indeed proved that the motion round the focus of the ellipse in which the sun lies is not of this description. One of his most important discoveries even related to the fact that at some parts of its orbit a planet swings around the sun with greater angular velocity than at others. But it so happens that in elliptic tracks which differ but little from circles, as is the case with all the more important planetary orbits, the motion round the empty focus of the ellipse is very nearly uniform. It seemed natural to assume, that this was exactly the case, in which event each of the two foci of the ellipse would have had a special significance in relation to the movement of the planet. The youthful Halley, however, demonstrated that so far as the empty focus was concerned, the movement of the planet around it, though so nearly uniform, was still not exactly so, and at the age of nineteen, he published a treatise on the subject which at once placed him in the foremost rank amongst theoretical astronomers.

But Halley had no intention of being merely an astronomer with his pen. He longed to engage in the practical work of observing. He saw that the progress of exact astronomy must depend largely on the determination of the positions of the stars with all attainable accuracy. He accordingly determined to take up this branch of work, which had been so successfully initiated by Tycho Brahe.

At the present day, astronomers of the great national observatories are assiduously engaged in the determination of the places of the stars. A knowledge of the exact positions of these bodies is indeed of the most fundamental importance, not alone for the purposes of scientific astronomy, but also for navigation and for extensive operations of surveying in which ac-curacy is desired. The fact that Halley determined to concentrate himself on this work shows clearly the scientific acumen of the young astronomer.

Halley, however, found that Hevelius, at Dantzig, and Flamsteed, the Astronomer Royal at Greenwich, were both engaged on work of this character. He accordingly determined to direct his energies in a way that he thought would be more useful to science. He resigned to the two astronomers whom I have named the investigation of the stars in the northern hemisphere, and he sought for himself a field hitherto almost entirely unworked. He determined to go to the southern hemisphere, there to measure and survey those stars which were invisible in Europe, so that his work should supplement the labours of the northern astronomers, and that the joint result of his labours and of theirs might be a complete survey of the most important stars on the surface of the heavens.

In these days, after so many ardent students everywhere have devoted themselves to the study of Nature, it seems difficult for a beginner to find a virgin territory in which to commence his explorations. Halley may, however, be said to have enjoyed the privilege of commencing to work in a magnificent region, the contents of which were previously almost entirely unknown. Indeed none of the stars which were so situated as to have been invisible from

Tycho Brahe's observatory at Uraniborg, in Denmark, could be said to have been properly observed. There was, no doubt, a rumour that a Dutchman had observed southern stars from the island of Sumatra, and certain stars were indicated in the southern heavens on a celestial globe. On examination, however, Halley found that no reliance could be placed on the results which had been obtained, so that practically the field before him may be said to have been unworked.

At the age of twenty, without having even waited to take that degree at the university which the authorities would have been glad to confer on so promising an undergraduate, this ardent student of Nature sought his father's permission to go to the southern hemisphere for the purpose of studying the stars which lie around the southern pole. His father possessed the necessary means, and he had likewise the sagacity to encourage the young astronomer. He was indeed most anxious to make everything as easy as possible for so hopeful a son. He provided him with an allowance of £300 a year, which was regarded as a very munificent provision in those days. Halley was also furnished with letters of recommendation from King Charles II., as well as from the directors of the East India Company. He accordingly set sail with his instruments in the year 1676, in one of the East India Company's ships, for the island of St. Helena, which he had selected as the scene of his labours.

After an uneventful voyage of three months, the astronomer landed on St. Helena, with his sextant of 5 feet radius, and a telescope 24 feet long, and forthwith plunged with ardour into his investigation of the southern skies. He met, however, with one very considerable disappointment. The climate of this island had been represented to him as most favourable for astronomical observation; but instead of the pure blue skies he had been led to expect, he found that they were almost always more or less clouded, and that rain was frequent, so that his observations were very much interrupted. On this account he only remained at St. Helena for a single year, having, during that time, and in spite of many difficulties, accomplished a piece of work which earned for him the title of "our southern Tycho. "Thus did Halley establish his fame as an astronomer on the same lonely rock in mid-Atlantic, which nearly a century and a-half later became the scene of Napoleon's imprisonment, when his star, in which he believed so firmly, had irretrievably set.

On his return to England, Halley prepared a map which showed the result of his labours, and he presented it to the king, in 1677. Like his great predecessor Tycho, Halley did not altogether disdain the arts of the courtier, for he endeavoured to squeeze a new constellation into the group around the southern pole which he styled "The Royal Oak," adding a description to the effect that the incidents of which "The Royal Oak" was a symbol were of sufficient importance to be inscribed on the surface of the heavens.

There is reason to think that Charles II. duly appreciated the scientific renown which one of his subjects had achieved, and it was probably through the influence of the king that Halley was made a Master of Arts at Oxford on November 18th, 1678. Special reference was made on the occasion to his observations at St. Helena, as evidence of unusual attainments in mathematics and astronomy. This degree was no small honour to such a young man, who, as we have seen, quitted his university before he had the opportunity of graduating in the ordinary manner.

On November 30th, in the same year, the astronomer received a further distinction in being elected a Fellow of the Royal Society. From this time forward he took a most active part in the affairs of the

Society, and the numerous papers which he read before it form a very valuable part of that notable series of volumes known as the "Philosophical Transactions." He was subsequently elected to the important office of secretary to the Royal Society, and he discharged the duties of his post until his appointment to Greenwich necessitated his resignation.

Within a year of Halley's election as a Fellow of the Royal Society, he was chosen by the Society to represent them in a discussion which had arisen with Hevelius. The nature of this discussion, or rather the fact that any discussion should have been necessary, may seem strange to modern astronomers, for the point is one on which it would now seem impossible for there to be any difference of opinion. We must, however, remember that the days of Halley were, comparatively speaking, the days of infancy as regards the art of astronomical observation, and issues that now seem obvious were often, in those early times, the occasions of grave and anxious consideration. The particular question on which Halley had to represent the Royal Society may be simply stated. When Tycho Brahe made his memorable investigations into the places of the stars, he had no telescopes to help him. The famous instruments at Uraniborg were merely provided with sights, by which the telescope was pointed to a star on the same principle as a rifle is sighted for a target. Shortly after Tycho's time, Galileo invented the telescope. Of course every one admitted at once the extraordinary advantages which the telescope had to offer, so far as the mere question of the visibility of objects was concerned. But the bearing of Galileo's invention upon what we may describe as the measuring part of astronomy was not so immediately obvious. If a star be visible to the unaided eye, we can determine its place by such instruments as those which Tycho used, in which no telescope is employed. We can, however, also avail ourselves of an instrument in which we view the star not directly but through the intervention of the telescope. Can the place of the star be determined more accurately by the latter method than it can when the telescope is dispensed with? With our present knowledge, of course, there is no doubt about the answer; every one conversant with instruments knows that we can determine the place of a star far more accurately with the telescope than is possible by any mere sighting apparatus. In fact an observer would be as likely to make an error of a minute with the sighting apparatus in Tycho's instrument, as he would be to make an error of a second with the modern telescope, or, to express the matter somewhat differently, we may say, speaking quite generally, that the telescopic method of determining the places of the stars does not lead to errors more than one-sixtieth part as great as the errors which are unavoidable when we make use of Tycho's method.

But though this is so apparent to the modern astronomer, it was not at all apparent in the days of Halley, and accordingly he was sent off to discuss the question with the Continental astronomers. Hevelius, as the representative of the older method, which Tycho had employed with such success, maintained that an instrument could be pointed more accurately at a star by the use of sights than by the use of a telescope, and vigorously disputed the claims put forward by those who believed that the latter method was the more suitable. On May 14th, 1679, Halley started for Dantzig, and the energetic character of the man may be judged from the fact that on the very night of his arrival he commenced to make the necessary observations. In those days astronomical telescopes had only obtained a fractional part of the perfection possessed by the instruments in our modern observatories, and therefore it may not be surprising that the results of the trial were not immediately conclusive.

- ROBERT STAWELL BALL

Halley appears to have devoted much time to the investigation; indeed, he remained at Dantzig for more than a twelvemonth. On his return to England, he spoke highly of the skill which Hevelius exhibited in the use of his antiquated methods, but Halley was nevertheless too sagacious an observer to be shaken in his preference for the telescopic method of observation.

The next year we find our young astronomer starting for a Continental tour, and we, who complain if the Channel passage lasts more than an hour or two, may note Halley's remark in writing to Hooke on June 15th, 1680: "Having fallen in with bad weather we took forty hours in the journey from Dover to Calais." The scientific distinction which he had already attained was such that he was received in Paris with marked attention. A great deal of his time seems to have been passed in the Paris observatory, where Cassini, the presiding genius, himself an astronomer of well-deserved repute, had extended a hearty welcome to his English visitor. They made observations together of the place of the splendid comet which was then attracting universal attention, and Halley found the work thus done of much use when he subsequently came to investigate the path pursued by this body. Halley was wise enough to spare no pains to derive all possible advantages from his intercourse with the distinguished *savants* of the French capital. In the further progress of his tour he visited the principal cities of the Continent, leaving behind him everywhere the memory of an amiable disposition and of a rare intelligence.

After Halley's return to England, in 1682, he married a young lady named Mary Tooke, with whom he lived happily, till her death fifty-five years later. On his marriage, he took up his abode in Islington, where he erected his instruments and recommenced his observations.

- ROBERT STAWELL BALL

It has often been the good fortune of astronomers to render practical services to humanity by their investigations, and Halley's achievements in this respect deserve to be noted. A few years after he had settled in England, he published an important paper on the variation of the magnetic compass, for so the departure of the needle from the true north is termed. This subject had indeed early engaged his attention, and he continued to feel much interest in it up to the end of his life. With respect to his labours in this direction, Sir John Herschel says: "To Halley we owe the first appreciation of the real complexity of the subject of magnetism. It is wonderful indeed, and a striking proof of the penetration and sagacity of this extraordinary man, that with his means of information he should have been able to draw such conclusions, and to take so large and comprehensive a view of the subject as he appears to have done." In 1692, Halley explained his theory of terrestrial magnetism, and begged captains of ships to take observations of the variations of the compass in all parts of the world, and to communicate them to the Royal Society, "in order that all the facts may be readily available to those who are hereafter to complete this difficult and complicated subject."

The extent to which Halley was in advance of his contemporaries, in the study of terrestrial magnetism, may be judged from the fact that the subject was scarcely touched after his time till the year 1811. The interest which he felt in it was not of a merely theoretical kind, nor was it one which could be cultivated in an easy-chair. Like all true investigators, he longed to submit his theory to the test of experiment, and for that purpose Halley determined to observe the magnetic variation for himself. He procured from King William III. the command of a vessel called the *Paramour Pink*, with which he started for the South Seas in 1694. This particular enterprise was not, however, successful; for, on crossing the line, some of his men fell sick and one of his lieutenants mutinied, so that he was obliged to return the following year with his mission unaccomplished. The government cashiered the lieutenant, and Halley having procured a second smaller vessel to accompany the *Paramour Pink*, started once more in September, 1699. He traversed the Atlantic to the 52nd degree of southern latitude, beyond which his further advance was stopped. "In these latitudes," he writes to say, "we fell in with great islands of ice of so incredible height and magnitude, that I scarce dare write my thoughts of it."

On his return in 1700, Halley published a general chart, showing the variation of the compass at the different places which he had visited. On these charts he set down lines connecting those localities at which the magnetic variation was identical. He thus set an example of the graphic representation of large masses of complex facts, in such a manner as to appeal at once to the eye, a method of which we make many applications in the present day.

But probably the greatest service which Halley ever rendered to human knowledge was the share which he took in bringing Newton's "Principia" before the world. In fact, as Dr. Glaisher, writing in 1888, has truly remarked, "but for Halley the 'Principia' would not have existed."

It was a visit from Halley in the year 1684 which seems to have first suggested to Newton the idea of publishing the results of his investigations on gravitation. Halley, and other scientific contemporaries, had no doubt some faint glimmering of the great truth which only Newton's genius was able fully to reveal. Halley had indeed shown how, on the assumptions that the planets move in circular orbits round the sun, and that the squares of their periodic times are proportional to the cubes of their mean distances, it may be proved that the force acting on each planet must vary inversely as the square of its distance from the sun. Since, however, each of the planets actually moves in an ellipse, and therefore, at continually varying distances from the sun, it becomes a much more difficult matter to account mathematically for the body's motions on the supposition that the attractive force varies inversely as the square of the distance. This was the question with which Halley found himself confronted, but which his mathematical abilities were not adequate to solve. It would seem that both Hooke and Sir Christopher Wren were interested in the same problem; in fact, the former claimed to have arrived at a solution, but declined to make known his results, giving as an excuse his desire that others having tried and failed might learn to value his achievements all the more. Halley, however, confessed that his attempts at the solution were unsuccessful, and Wren, in order to encourage the other two philosophers to pursue the inquiry, offered to present a book of forty shillings' value to either of them who should in the space of two months bring him a convincing proof of it. Such was the value which Sir Christopher set on the Law of Gravitation, upon which the whole fabric of modern astronomy may be said to stand.

Finding himself unequal to the task, Halley went down to Cambridge to see Newton on the subject, and was delighted to learn that the great mathematician had already completed the investigation. He showed Halley that the motions of all the planets could be completely accounted for on the hypothesis of a force of attraction directed towards the sun, which varies inversely as the square of the distance from that body.

Halley had the genius to perceive the tremendous importance of Newton's researches, and he ceased not to urge upon the recluse man of science the necessity for giving his new discoveries publication. He paid another visit to Cambridge with the object of learning more with regard to the mathematical methods which had already conducted Newton to such sublime truths, and he again encouraged the latter both to pursue his investigations, and to give some account of them to the world. In December of the same year Halley had the gratification of announcing to the Royal Society that Newton had promised to send that body a paper containing his researches on Gravitation.

It seems that at this epoch the finances of the Royal Society were at a very low ebb. This impecuniosity was due to the fact that a book by Willoughby, entitled "De Historia Piscium," had been recently printed by the society at great expense. In fact, the coffers were so low that they had some difficulty in paying the salaries of their permanent officials. It appears that the public did not care about the history of fishes, or at all events the volume did not meet with the ready demand which was expected for it. Indeed, it has been recorded that when Halley had undertaken to measure the length of a degree of the earth's surface, at the request of the Royal Society, it was ordered that his expenses be defrayed either in £50 sterling, or in fifty books of fishes. Thus it happened that on June 2nd, the Council, after due consideration of ways and means in connection with the issue of the "Principia," ordered that Halley "should undertake the business of looking after the book and printing it at his own charge," which he engaged to do.

It was, as we have elsewhere mentioned, characteristic of Newton that he detested controversies, and he was, in fact, inclined to suppress the third book of the "Principia" altogether rather than have any conflict with Hooke with respect to the discoveries there enunciated. He also thought of changing the name of the work to *De Motu Corporum Libri Duo*, but upon second thoughts, he retained the original title, remarking, as he wrote to Halley, "It will help the sale of the book, which I ought not to diminish, now it is yours," a sentence which shows conclusively, if further proof were necessary, that Halley had assumed the responsibility of its publication.

Halley spared no pains in pushing forward the publication of his illustrious friend's great work, so that in the same year he was in a position to present a complete copy to King James II., with a proper discourse of his own. Halley also wrote a set of Latin hexameters in praise of Newton's genius, which he printed at the beginning of the work. The last line of this specimen of Halley's poetic muse may be thus rendered: "Nor mortals nearer may approach the gods."

The intimate friendship between the two greatest astronomers of the time continued without interruption till the death of Newton. It has, indeed, been alleged that some serious cause of estrangement arose between them. There is, however, no satisfactory ground for this statement; indeed, it may be regarded as effectually disposed of by the fact that, in the year 1727, Halley took up the defence of his friend, and wrote two learned papers in support of Newton's "System of Chronology," which had been seriously attacked by a certain ecclesiastic. It is quite evident to anyone who has studied these papers that Halley's friendship for Newton was as ardent as ever.

The generous zeal with which Halley adopted and defended the doctrines of Newton with regard to the movements of the celestial bodies was presently rewarded by a brilliant discovery, which has more than any of his other researches rendered his name a familiar one to astronomers. Newton, having explained the movements of the planets, was naturally led to turn his attention to comets. He perceived that their journeyings could be completely accounted for as consequences of the attraction of the sun, and he laid down the principles by which the orbit of a comet could be determined, provided that observations of its positions were obtained at three different dates. The importance of these principles was by no one more quickly recognised than by Halley, who saw at once that it provided the means of detecting something like order in the movements of these strange wanderers. The doctrine of Gravitation seemed to show that just as the planets revolved around the sun in ellipses, so also must the comets. The orbit, however, in the case of a comet, is so extremely elongated that the very small part of the elliptic path within which the comet is both near enough and bright enough to be seen from the earth, is indistinguishable from a parabola. Applying these principles, Halley thought it would be instructive to study the movements of certain bright comets, concerning which reliable observations could be obtained. At the expense of much labour, he laid down the paths pursued by twenty-four of these bodies, which had appeared between the years 1337 and 1698. Amongst them he noticed three, which followed tracks so closely resembling each other, that he was led to conclude the so-called three comets could only have been three different appearances of the same body. The first of these occurred in 1531, the second was seen by Kepler in 1607, and the third by Halley himself in 1682. These dates suggested that the observed phenomena might be due to the successive returns of one and the same comet after intervals of seventy-five or seventy-six years. On the further examination of ancient records, Halley found that a comet had been seen in the year 1456, a date, it will be observed, seventy-five years before 1531. Another had been observed seventy-six years earlier than 1456, viz., in 1380, and another seventy-five years before that, in 1305.

As Halley thus found that a comet had been recorded on several occasions at intervals of seventy-five or seventy-six years, he was led to the conclusion that these several apparitions related to one and the same object, which was an obedient vassal of the sun, performing an eccentric journey round that luminary in a period of seventy-five or seventy-six years. To realise the importance of this discovery, it should be remembered that before Halley's time a comet, if not regarded merely as a sign of divine displeasure, or as an omen of intending disaster, had at least been regarded as a chance visitor to the solar system, arriving no one knew whence, and going no one knew whither.

A supreme test remained to be applied to Halley's theory. The question arose as to the date at which this comet would be seen again. We must observe that the question was complicated by the fact that the body, in the course of its voyage around the sun, was exposed to the incessant disturbing action produced by the attraction of the several planets. The comet, therefore, does not describe a simple ellipse as it would do if the attraction of the sun were the only force by which its movement were controlled. Each of the planets solicits the comet to depart from its track, and though the amount of these attractions may be insignificant in comparison with the supreme controlling force of the sun, yet the departure from the ellipse is quite sufficient to produce appreciable irregularities in the comet's movement. At the time when Halley lived, no means existed of calculating with precision the effect of the disturbance a comet might experience from the action of the different planets. Halley exhibited his usual astronomical sagacity in deciding that Jupiter would retard the return of the comet to some extent. Had it not been for this disturbance the comet would apparently have been due in 1757 or early in 1758. But the attraction of the great planet would cause delay, so that Halley assigned, for the date of its re-appearance, either the end of 1758 or the beginning of 1759. Halley knew that he could not himself live to witness the fulfilment of his prediction, but he says: "If it should return, according to our predictions, about the year 1758, impartial posterity will not refuse to acknowledge that this was first discovered by an Englishman." This was, indeed, a remarkable prediction of an event to occur fifty-three years after it had been uttered. The way in which it was fulfilled forms one of the most striking episodes in the history of astronomy. The comet was first seen on Christmas Day, 1758, and passed through its nearest point to the sun on March 13th, 1759. Halley had then been lying in his grave for seventeen years, yet the verification of his prophecy reflects a glory on his name, which will cause it to live forever in the annals of astronomy. The comet paid a subsequent visit in 1835, and its next appearance is due about 1910.

Halley next entered upon a labour which, if less striking to the imagination than his discoveries with regard to comets, is still of inestimable value in astronomy. He undertook a series of investigations with the object of improving our knowledge of the movements of the planets. This task was practically finished in 1719, though the results of it were not published until after his death in 1749. In the course of it he was led to investigate closely the motion of Venus, and thus he came to recognise for the first time the peculiar importance which attaches to the phenomenon of the transit of this planet across the sun. Halley saw that the transit, which was to take place in the year 1761, would afford a favourable opportunity for determining the distance of the sun, and thus learning the scale of the solar system. He predicted the circumstances of the phenomenon with an astonishing degree of accuracy, considering his means of information, and it is unquestionably to the exertions of Halley in urging the importance of the matter upon astronomers that we owe the unexampled degree of interest taken in the event, and the energy which scientific men exhibited in observing it. The illustrious astronomer had no hope of being himself a witness of the event, for it could not happen till many years after his death. This did not, however, diminish his anxiety to impress upon those who would then be alive, the importance of the occurrence, nor did it lead him to neglect anything which might contribute to the success of the observations. As we now know, Halley rather over-estimated the value of the transit of Venus as a means of determining the solar distance. The fact is that the circumstances are such that the observation of the time of contact between the edge of the planet and the edge of the sun cannot be made with the accuracy which he had expected.

In 1691, Halley became a candidate for the Savilian Professorship of Astronomy at Oxford. He was not, however, successful, for his candidature was opposed by Flamsteed, the Astronomer Royal of the time, and another was appointed. He received some consolation for this particular disappointment by the fact that, in 1696, owing to Newton's friendly influence, he was appointed deputy Controller of the Mint at Chester, an office which he did not retain for long, as it was abolished two years later. At last, in 1703, he received what he had before vainly sought, and he was appointed to the Savilian chair.

His observations of the eclipse of the sun, which occurred in 1715, added greatly to Halley's reputation. This phenomenon excited special attention, inasmuch as it was the first total eclipse of the sun which had been visible in London since the year 1140. Halley undertook the necessary calculations, and predicted the various circumstances with a far higher degree of precision than the official announcement. He himself observed the phenomenon from the Royal Society's rooms, and he minutely describes the outer atmosphere of the sun, now known as the corona; without, however, offering an opinion as to whether it was a solar or a lunar appendage.

At last Halley was called to the dignified office which he of all men was most competent to fill. On February 9th, 1720, he was appointed Astronomer Royal in succession to Flamsteed. He found things at the Royal Observatory in a most unsatisfactory state. Indeed, there were no instruments, nor anything else that was movable; for such things, being the property of Flamsteed, had been removed by his widow, and though Halley attempted to purchase from that lady some of the instruments which his predecessor had employed, the unhappy personal differences which had existed between him and Flamsteed, and which, as we have already seen, prevented his election as Savilian Professor of Astronomy, proved a bar to the negotiation. Greenwich Observatory wore a very different appearance in those days, from that which the modern visitor, who is fortunate enough to gain admission, may now behold. Not only did Halley find it bereft of instruments, we learn besides that he had no assistants, and was obliged to transact the whole business of the establishment single-handed.

In 1721, however, he obtained a grant of £500 from the Board of Ordnance, and accordingly a transit instrument was erected in the same year. Sometime afterwards he procured an eight-foot quadrant, and with these instruments, at the age of sixty-four, he commenced a series of observations on the moon. He intended, if his life was spared, to continue his observations for a period of eighteen years, this being, as astronomers know, a very important cycle in connection with lunar movements. The special object of this vast undertaking was to improve the theory of the moon's motion, so that it might serve more accurately to determine longitudes at sea. This self-imposed task Halley lived to carry to a successful termination, and the tables deduced from his observations, and published after his death, were adopted almost universally by astronomers, those of the French nation being the only exception.

Throughout his life Halley had been singularly free from illness of every kind, but in 1737 he had a stroke of paralysis. Notwithstanding this, however, he worked diligently at his telescope till 1739, after which his health began rapidly to give way. He died on January 14th, 1742, in the eighty-sixth year of his age, retaining his mental faculties to the end. He was buried in the cemetery of the church of Lee in Kent, in the same grave as his wife, who had died five years previously. We are informed by Admiral Smyth that Pond, a later Astronomer Royal, was afterwards laid in the same tomb.

Halley's disposition seems to have been generous and candid, and wholly free from anything like jealousy or rancour. In person he was rather above the middle height, and slight in build; his complexion was fair, and he is said to have always spoken, as well as acted, with uncommon sprightliness. In the *eloge* pronounced upon him at the Paris *Acadernie des Sciences*, of which Halley had been made a member in 1719, it was said, "he possessed all the qualifications which were necessary to please princes who were desirous of instruction, with a great extent of knowledge and a constant presence of mind; his answers were ready, and at the same time pertinent, judicious, polite and sincere."

- ROBERT STAWELL BALL

Thus we find that Peter the Great was one of his most ardent admirers. He consulted the astronomer on matters connected with shipbuilding, and invited him to his own table. But Halley possessed nobler qualifications than the capacity of pleasing princes.

Greenwich Observatory in Halley's Time.

He was able to excite and to retain the love and admiration of his equals. This was due to the warmth of his attachments, the unselfishness of his devotion to his friends, and to a vein of gaiety and good humour which pervaded all his conversation.

COMPLIMENTARY MATERIAL

A BRIEF BIOGRAPHICAL SKETCH

EDMOND HALLEY

- ROBERT STAWELL BALL

Edmond Halley

(November 8, 1656 – January 25, 1742)

COMPLIMENTARY MATERIAL

A BRIEF BIOGRAPHICAL SKETCH

EDMOND HALLEY

Edmond Halley (November 8, 1656 – January 25, 1742) was an English astronomer, geophysicist, mathematician, meteorologist, and physicist who is best known for computing the orbit of Halley's Comet. He was the second Astronomer Royal in Britain, succeeding John Flamsteed.

Early Life

Halley was born in Haggerston, in east London. His father, Edmond Halley Sr., came from a Derbyshire family and was a wealthy soap-maker in London. As a child, Halley was very interested in mathematics. He studied at St Paul's School, and from 1673 at The Queen's College, Oxford. While still an undergraduate, Halley published papers on the Solar System and sunspots.

Career

Publications and Inventions

Halley became an assistant to John Flamsteed, the Astronomer Royal at the Greenwich Observatory, in 1675, and among other things, had the job of assigning what is now called Flamsteed numbers to stars.

In 1676, Halley visited the south Atlantic island of Saint Helena and set up an observatory with a large sextant with telescopic sights to catalogue the stars of the southern hemisphere. While there he observed a transit of Mercury, and realised that a similar transit of Venus could be used to determine the absolute size of the Solar System. He returned to England in May 1678. In the following year he went to Danzig (Gdańsk) on behalf of the Royal Society to help resolve a dispute. Because astronomer Johannes Hevelius did not use a telescope, his observations had been questioned by Robert Hooke. Halley stayed with Hevelius and he observed and verified the quality of Hevelius' observations. In 1679 Halley published the results from his observations on St. Helena as Catalogus Stellarum Australium which included details of 341 southern stars. These additions to contemporary star maps earned him comparison with Tycho Brahe: e.g. "the southern Tycho" as described by Flamsteed. Halley was awarded his M.A. degree at Oxford and elected as a Fellow of the Royal Society at the age of 22.

In 1686, Halley published the second part of the results from his Helenian expedition, being a paper and chart on trade winds and monsoons. The symbols he used to represent trailing winds still exist in most modern day weather chart representations. In this article he identified solar heating as the cause of atmospheric motions. He also established the relationship between barometric pressure and height above sea level. His charts were an important contribution to the emerging field of information visualisation.

Bust of Halley
(Royal Observatory, Greenwich)

Halley spent most of his time on lunar observations, but was also interested in the problems of gravity. One problem that attracted his attention was the proof of Kepler's laws of planetary motion. In August 1684, he went to Cambridge to discuss this with Isaac Newton, much as John Flamsteed had done four years earlier, only to find that Newton had solved the problem, at the instigation of Flamsteed with regard to the orbit of comet Kirch, without publishing the solution. Halley asked to see the calculations and was told by Newton that he could not find them, but promised to redo them and send them on later, which he eventually did, in a short treatise entitled, On the motion of bodies in an orbit. Halley recognised the importance of the work and returned to Cambridge to arrange its publication with Newton, who instead went on to expand it into his Philosophiæ Naturalis Principia Mathematica published at Halley's expense in 1687. Halley's first calculations with comets were thereby for the orbit of comet Kirch, based on Flamsteed's observations in 1680-1. Although he was to accurately calculate the orbit of the comet of 1682, he was inaccurate in his calculations of the orbit of comet Kirch. They indicated a periodicity of 575 years, thus appearing in the years 531 and 1106, and presumably heralding the death of Julius Caesar in a like fashion in −44 (45 BCE). It is now known to have an orbital period of circa 10,000 years.

In 1691, Halley built a diving bell, a device in which the atmosphere was replenished by way of weighted barrels of air sent down from the surface. In a demonstration, Halley and five companions dived to 60 feet (18 m) in the River Thames, and remained there for over an hour and a half. Halley's bell was of little use for practical salvage work, as it was very heavy, but he made improvements to it over time, later extending his underwater exposure time to over 4 hours. Halley suffered one of the earliest recorded cases of middle ear barotrauma. That same year, at a meeting of the Royal Society, Halley introduced a rudimentary working model of a magnetic compass using a liquid-filled housing to damp the swing and wobble of the magnetised needle.

In 1691 Halley sought the post of Savilian Professor of Astronomy at Oxford, but, due to being accused of atheism, was opposed by the Archbishop of Canterbury, John Tillotson, and Bishop Stillingfleet. The post went instead to David Gregory, who had the support of Isaac Newton.

In 1692, Halley put forth the idea of a hollow Earth consisting of a shell about 500 miles (800 km) thick, two inner concentric shells and an innermost core. He suggested that atmospheres separated these shells, and that each shell had its own magnetic poles, with each sphere rotating at a different speed. Halley proposed this scheme to explain anomalous compass readings. He envisaged each inner region as having an atmosphere and being luminous (and possibly inhabited), and speculated that escaping gas caused the Aurora Borealis.

In 1693 Halley published an article on life

annuities, which featured an analysis of age-at-death on the basis of the Breslau statistics Caspar Neumann had been able to provide. This article allowed the British government to sell life annuities at an appropriate price based on the age of the purchaser. Halley's work strongly influenced the development of actuarial science. The construction of the life-table for Breslau, which followed more primitive work by John Graunt, is now seen as a major event in the history of demography.

The Royal Society censured Halley for suggesting in 1694 that the story of Noah's flood might be an account of a cometary impact.

Exploration Years

In 1698, Halley was given command of the Paramour, a 52 feet (16 m) pink, so that he could carry out investigations in the South Atlantic into the laws governing the variation of the compass. On 19 August 1698, he took command of the ship and, in November 1698, sailed on what was the first purely scientific voyage by an English naval vessel. Unfortunately problems of insubordination arose over questions of Halley's competence to command a vessel. Halley returned the ship to England to proceed against officers in July 1699. The result was a mild rebuke for his men, and dissatisfaction for Halley, who felt the court had been too lenient. Halley thereafter received a temporary commission as a Captain in the Royal Navy, recommissioned the Paramour on 24 August 1699 and sailed again in September 1699 to make extensive observations on the conditions of terrestrial magnetism. This task he accomplished in a second Atlantic voyage which lasted until 6 September 1700, and extended from 52 degrees north to 52 degrees south. The results were published in General Chart of the Variation of the Compass (1701). This was the first such chart to be published and the first on which isogonic, or Halleyan, lines appeared.

The preface to Awnsham and John Churchill's collection of voyages and travels (1704), supposedly written by John Locke or by Halley.

"Natural and moral history is embellished with the most beneficial increase of so many thousands of plants it had never before received, so many drugs and spices, such unaccountable diversity. Trade is raised to highest pitch, and this not in a niggard and scanty manner as when the Venetians served all Europe ... the empire of Europe is now extended to the utmost bounds of the Earth."

In November 1703, Halley was appointed Savilian Professor of Geometry at the University of Oxford, his theological enemies, John Tillotson and Bishop Stillingfleet having died, and received an honorary degree of doctor of laws in 1710. In 1705, applying historical astronomy methods, he published Synopsis Astronomia Cometicae, which stated his belief that the comet sightings of 1456, 1531, 1607, and 1682 were of the same comet, which he predicted would return in 1758. Halley did not live to witness the comet's return, but when it did, the comet became generally known as Halley's Comet.

By 1706 Halley had learned Arabic and completed the translation started by Edward Bernard of Books V-VII of Apollonius's Conics from copies found at Leiden and the Bodleian Library at Oxford. He also completed a new translation of the first four books from the original Greek that had been started by the late David Gregory. He published these along with his own reconstruction of Book VIII in the first complete Latin edition in 1710.

In 1716, Halley suggested a high-precision measurement of the distance between the Earth and the Sun by timing the transit of Venus. In doing so, he was following the method described by James Gregory in Optica Promota (in which the design of the Gregorian telescope is also described). It is reasonable to assume Halley possessed and had read this book given that the Gregorian design was the principal telescope design used in astronomy in Halley's day. It is not to Halley's credit that he failed to acknowledge Gregory's priority in this matter. In 1718 he discovered the proper motion of the "fixed" stars by comparing his astrometric measurements with those given in Ptolemy's Almagest. Arcturus and Sirius were two noted to have moved significantly, the latter having progressed 30 arc minutes (about the diameter of the moon) southwards in 1800 years.

In 1720, together with his friend the antiquarian William Stukeley, Halley participated in the first attempt to scientifically date Stonehenge. Assuming that the monument had been laid out using a magnetic compass, Stukeley and Halley attempted to calculate the perceived deviation introducing corrections from existing magnetic records, and suggested three dates (460 BC, 220 AD and 920 AD), the earliest being the one accepted. These dates were wrong by thousands of years, but the idea that scientific methods could be used to date ancient monuments was revolutionary in its day.

Halley's Grave

Edmund Halley at St Margaret's, Lee, Churchyard.

Halley succeeded John Flamsteed in 1720 as Astronomer Royal, a position Halley held until his death.

Halley died in 1742 at the age of 85. He was buried in the graveyard of the old church of St Margaret's, Lee (since rebuilt), at Lee Terrace, Blackheath. He was interred in the same vault as the Astronomer Royal John Pond; the unmarked grave of the Astronomer Royal Nathaniel Bliss is nearby.

His original tombstone was transferred by the Admiralty when the original Lee church was demolished and rebuilt – it can be seen today on the southern wall of the Camera Obscura at the Royal Observatory, Greenwich. His marked grave can be seen at St Margaret's Church, Lee Terrace.

Plaque in South Cloister of Westminster Abbey

Halley's map of the path of the Solar eclipse of 3 May 1715 across England

Personal Life

Halley married Mary Tooke in 1682 and settled in Islington. The couple had three children.

Named after Edmond Halley

Halley's map of the path of the Solar eclipse of 3 May 1715 across England.

- Halley's Comet (orbital period (approximately) 75 years)
- Halley (lunar crater)
- Halley (Martian crater)
- Halley Research Station, Antarctica
- Halley's method, for the numerical solution of equations
- Halley Street, in Blackburn, Victoria, Australia
- Edmund Halley Road, Oxford Science Park, Oxford, OX4 4DQ UK
- Edmund Halley Drive, Reston, Virginia, USA
- Halley Ward, surgical ward at Homerton Hospital East London

- Halley's Mount, Saint Helena (680m high)

- Halley Drive, Hackensack, NJ. Intersects with Comet Way on the campus of Hackensack High School, The home of The Comets

- Rue Edmund Halley, Avignon, France

- The Edmund Halley, A JD Wetherspoon pub in Lee Green, London

- Sir Edmond Halley's Restaurant & Freehouse, a pub in Charlotte, North Carolina

ABOUT THE AUTHOR

ROBERT STAWELL BALL
(July 1, 1840 – November 25, 1913)

Sir Robert Stawell Ball was an Irish astronomer who founded the screw theory.

He was the son of naturalist Robert Ball and Amelia Gresley Hellicar. He was born in Dublin.

Ball worked for Lord Rosse from 1865 to 1867. In 1867 he became Professor of Applied Mathematics at the Royal College of Science in Dublin. There he lectured on mechanics and published an elementary account of the science.

In 1874 Ball was appointed Royal Astronomer of Ireland and Andrews Professor of Astronomy in the University of Dublin at Dunsink Observatory.

Ball contributed to the science of kinematics by delineating the screw displacement:

When Ball and the screw theorists speak of screws they no longer mean actual cylindrical objects with helical threads cut into them but the possible motion of any body whatsoever, including that of the screw independently of the nut.

Ball's treatise *The Theory of Screws* (1876) is now in the public domain. His work on screw dynamics earned him in 1879 the

Cunningham Medal of the Royal Irish Academy.

In 1882 Popular Science Monthly carried his article "*A Glimpse through the Corridors of Time*". The following year it carried his two-part article on "*The Boundaries of Astronomy*".

Ball expounded the tides in *Time and Tide*: a Romance of the Moon In 1892 he was appointed Lowndean Professor of Astronomy and Geometry at Cambridge University at the same time becoming director of the Cambridge Observatory. He was a fellow of King's College, Cambridge.

In 1900 Cambridge University Press published *A Treatise on the Theory of Screws*. That year he also published *The Story of the Heavens*. Much in the limelight, he stood as President of the Quaternion Society. He was also President of the Mathematical Association in 1900.

In 1908 he published *A Treatise on Spherical Astronomy*, which is a textbook on astronomy starting from spherical trigonometry and the celestial sphere, considering atmospheric refraction and aberration of light, and introducing basic use of a generalised instrument.

His work *The Story of the Heavens* is mentioned in the "Ithaka" chapter of Ulysses. His lectures, articles and books (e.g. *Starland* and *The Story of the Heavens*) were mostly popular and simple in style.

He died in Cambridge and was buried at the Parish of the Ascension Burial Ground in Cambridge, with his wife Lady Francis Elizabeth Ball. Their children were: Frances Amelia, Robert Steele, William Valentine (later Sir), Mary Agnetta, Charles Rowan Hamilton, and Randall Gresley (later Colonel).

LECTURES

Ball became celebrated for his popular lectures on science. He gave an estimated 2500 lectures between 1875 and 1910 in towns and cities across Britain and Ireland. In 1892, 1898 and 1900 he was invited to deliver the Royal Institution Christmas Lecture. *Astronomy; Astronomy* and *Great Chapters from the Book of Nature.*

FEEDBACK

Now that you have read the book ...

Was it interesting?

Did you enjoy what you wanted to read?
Was there any room for improvement?

Let us know at:

http://www.diamondbooks.ca/feedback

Your feedback is highly appreciated.

Thank you!

NOTES

NOTES

NOTES

NOTES

NOTES

NOTES

Made in United States
Cleveland, OH
06 November 2026

47647264R00055